Abdelhafid Mimouni

The essence of synchrotron radiation

Abdelhafid Mimouni

The essence of synchrotron radiation

ScienciaScripts

Imprint
Any brand names and product names mentioned in this book are subject to trademark, brand or patent protection and are trademarks or registered trademarks of their respective holders. The use of brand names, product names, common names, trade names, product descriptions etc. even without a particular marking in this work is in no way to be construed to mean that such names may be regarded as unrestricted in respect of trademark and brand protection legislation and could thus be used by anyone.

Cover image: www.ingimage.com

This book is a translation from the original published under ISBN 978-620-6-72715-6.

Publisher:
Sciencia Scripts
is a trademark of
Dodo Books Indian Ocean Ltd. and OmniScriptum S.R.L publishing group

120 High Road, East Finchley, London, N2 9ED, United Kingdom
Str. Armeneasca 28/1, office 1, Chisinau MD-2012, Republic of Moldova, Europe
Printed at: see last page
ISBN: 978-620-8-28847-1

The essence of synchrotron radiation

Author: Dr. Abdelhafid Mimouni: Independent researcher in bioinorganic chemistry, holder of a doctorate in chemistry from the University of Paris XII (1997) and a Diplôme des Études Approfondies en Systèmes Bioinorganiques from the University of Paris XI (93), a Licence and a Maîtrise in chemistry (91, 92).

Summary: Synchrotron radiation is an essential phenomenon in physics, based on Maxwell's equations. This book has examined the dynamics of electrons in a magnetic field, highlighting the importance of relativity in the emission of radiation. We have detailed the characteristics of synchrotron radiation, such as its spectrum, directionality and intensity, while presenting the equations needed to calculate the radiated power. The various applications of synchrotron radiation in scientific and industrial research were explored, particularly in the fields of biology, medicine and materials. Finally, technological advances in accelerators are opening up new research horizons, promising significant discoveries. The continuing importance of this study lies in its potential to revolutionise various scientific fields and foster technological innovation.

Book outline :

Introduction

Synchrotron radiation is a fascinating phenomenon that results from the acceleration of charged particles, particularly electrons, when they are subjected to intense magnetic fields. Discovered in the mid-20th century, synchrotron radiation was quickly recognised for its fundamental importance in particle physics and technological applications. The work of pioneers such as Gregory Breit, John A. Wheeler and Robert Wilson established the theoretical and experimental basis for this phenomenon, paving the way for significant advances in various scientific disciplines.

History of synchrotron radiation

Early studies (1930s)

The first studies on radiation from accelerated charges date back to the 1930s. Physicists such as **Gregory Breit** and **John A. Wheeler** examined the effects of electromagnetic fields on charged particles. Their work laid the theoretical foundations for understanding the interactions between particles and magnetic fields, paving the way for what would become the theory of synchrotron radiation.

Experimental discovery (1940s)

In the 1940s, the phenomenon was observed experimentally in particle accelerator installations. **Robert Wilson,** who was working on the first synchrotron at the University of Illinois, was among the first to document the radiation emitted by electrons circulating in a magnetic field. In 1947, experiments demonstrated that electrons emit intense light when accelerated in magnetic fields. This radiation, initially considered an artefact, was recognised for its importance and given the name "synchrotron radiation".

Theoretical and technological advances (1960s-1970s)

As particle accelerator technology advanced, physicists began to exploit synchrotron radiation as a source of intense light. In the 1960s, the construction of dedicated research synchrotrons, such as the **Stanford Synchrotron** and other similar facilities, led to the development of imaging and diffraction techniques that revolutionised many areas of science. The work of **John D. Jackson**, who formulated equations detailing the properties of synchrotron radiation, was also fundamental to the theoretical understanding of this phenomenon.

Modern applications (1980s to today)

From the 1980s onwards, the applications of synchrotron radiation multiplied, affecting a wide range of fields such as materials physics, structural biology and chemistry. Synchrotron facilities around the world, such as the **European Synchrotron Radiation Facility (ESRF)** and the **Advanced Photon Source (APS)**, were created to take advantage of the unique properties of synchrotron radiation. These facilities enable high-resolution studies of atomic and molecular structure, contributing to major advances in biomedical research, the development of new materials and the understanding of fundamental chemical processes.

Importance and applications in science and technology

Synchrotron radiation plays a crucial role in many fields of science, including physics, chemistry, biology and materials science. It is used in advanced imaging, X-ray diffraction and spectroscopy techniques, enabling researchers to explore the atomic and molecular structure of materials with unprecedented precision. In addition, industrial and medical applications, such as materials analysis, the development of new drugs and biomedical research, take advantage of the unique properties of synchrotron radiation.

Aims and structure of the book

This book aims to provide an in-depth understanding of synchrotron radiation, exploring its fundamental principles, emission mechanisms and applications. Using a methodical approach, we cover the basic concepts of electromagnetism and relativity, followed by an analysis of the movement of electrons and the consequences of their circulation in magnetic fields. The following chapters will detail the phenomenon of synchrotron radiation itself, as well as its applications in scientific research and industry.

The book is structured into several chapters, each dealing with a particular aspect of synchrotron radiation, from its theoretical foundations to its practical applications. We hope that this book will serve as a valuable reference for students, researchers and professionals wishing to gain a better understanding

of this phenomenon and its implications.

Chapter 1: Fundamentals of electromagnetism

1.1 Maxwell's equations

Maxwell's equations, formulated by the Scottish physicist James Clerk Maxwell in the 19th century, are the foundations of electromagnetism. They describe how electric and magnetic fields interact with electric charges and currents. There are four equations, which can be expressed in integral or differential form. Here is an overview of each of them:

1. **Gauss's law for electricity** :

$$\nabla \cdot E = \varepsilon_0 \rho$$

 This equation states that the divergence of the electric field E is proportional to the charge density ρ. It shows that electric charges create electric fields.

2. **Gauss's law of magnetism** :

 $$\nabla \cdot \mathbf{B} = 0$$

 This equation indicates that there are no magnetic monopolies; the magnetic field lines B are always closed.

3. **Faraday's Law** :

 $$\nabla \times \mathbf{E} = -\partial t \partial \mathbf{B}$$

It expresses that the temporal variation of a magnetic field induces an electric field, which is the basis of the operation of electric generators.

4. **Ampère-Maxwell's law** :

$$\nabla \times B = \mu_0 J + \mu_0 \varepsilon_0 \, (\partial E / \partial t)$$

This equation relates the magnetic field B to the current density J and to the time variation of the electric field E. It also introduces the concept of electric displacement, represented by the term $\boldsymbol{\mu_0 \varepsilon_0}$ $(\partial \mathbf{E}/\partial \mathbf{t})$, which makes it possible to take into account situations where electric fields vary over time, and therefore completes the classical Ampère's law.

Maxwell's equations show how electric and magnetic fields are interconnected and how they propagate in space, forming the basis of electrodynamics.

Interactions between electric and magnetic fields

Electric and magnetic fields interact dynamically. A static electric field can generate a changing magnetic field if the electric field changes over time, as described by Faraday's law. Similarly, a moving or changing magnetic field can induce an electric field, according to Ampère-Maxwell's law. This interaction is fundamental to understanding electromagnetic radiation, where variations in a field can lead to the propagation of electromagnetic waves.

1.2 Forces on charged particles

Charged particles, such as electrons, are influenced by electric and magnetic fields, resulting in forces acting on them. The resulting force on a charged particle is described by the Lorentz force.

Lorentz force and its application

The Lorentz force is given by the equation :

$$F=q(E+v\times B)$$

Where q is the charge of the particle, E is the electric field, v is the velocity of the particle and B is the magnetic field. This force is the sum of the effects of the electric and magnetic fields. The Lorentz force deflects charged particles when they move in a magnetic field, which is essential for the flow of electrons in accelerators and synchrotrons.

Relativistic effects

When a charged particle travels at speeds close to that of light, relativistic effects must be taken into account. The dynamics of relativistic particles is described by Einstein's theory of special relativity. The Lorentz factor γ is introduced to take account of the increase in effective mass and changes in velocity:

$$\gamma=1/(1-(v^2/c^2))^{-1/2}$$

At relativistic speeds, the Lorentz force must be reassessed, and the effects of kinetic energy become significant. This has profound implications for understanding synchrotron radiation, where electrons travelling at relativistic speeds emit intense radiation when deflected in a magnetic field.

Chapter 2: Electron motion

2.1. Electron acceleration

Electron acceleration is a key process in research and applications using synchrotron radiation. Electrons can be accelerated in a number of ways, including by electric or magnetic fields, and depending on the type of movement they make.

Types of acceleration (linear and circular)

1. **Linear acceleration** :
    - In a linear configuration, electrons are subjected to a constant electric field that drives them at increasing speeds. This typically occurs in linear particle accelerators (linacs) where electrons acquire energy through resonant cavities.
    - The equation of motion in this case is governed by :

F=qE

Where F is the force, q is the charge of the electron and E is the intensity of the electric field. The acceleration is therefore directly proportional to the applied field.

2. **Circular acceleration** :
    - In circular accelerators, such as synchrotrons, the electrons are held in orbit by a magnetic field. When they change direction, they

undergo a centripetal acceleration that keeps them on their circular trajectory.

- o This form of acceleration is crucial for synchrotron radiation, as centripetal acceleration is accompanied by radiation emission.

Equation of motion of an electron in a magnetic field

For an electron moving in a magnetic field, the Lorentz force describes the resulting acceleration:

$$F=q(E+v\times B)$$

When the electron is only under the effect of a magnetic field (i.e. E=0), the acceleration can be expressed as :

$$m(dv/dt)=q(v\times B)$$

where mmm is the mass of the electron. This equation shows that the force acting on the electron is always perpendicular to its velocity, resulting in circular motion.

2.2 Circulation of electrons in a storage ring

Storage rings are devices used to hold electrons at high energies in a circular trajectory, enabling continuous generation of synchrotron radiation. Understanding the geometric configuration and the forces at play is essential for analysing how they work.

Geometric configuration of the ring

A typical storage ring consists of a circular path into which electrons are injected and held in orbit by magnetic dipoles. These dipoles create a magnetic field that bends the trajectory of the electrons. Quadrupoles and other devices are also used to stabilise the orbit and compensate for deviation effects.

The dimensions of the ring and the strength of the magnetic field are carefully designed to ensure that electrons can be stored for extended periods while minimising energy losses due to radiation.

Balance between centripetal forces and Lorentz forces

For an electron to move around a storage ring, there must be a balance between the centripetal force needed to maintain its circular trajectory and the Lorentz force due to the magnetic field. This balance can be expressed as :

$$m v^2 / r = q v B$$

Where r is the radius of the circular trajectory. By rearranging this equation, we can express the speed of the electrons as a function of the parameters of the system:

$$v = q B r / m$$

This relationship is crucial to understanding the dynamics of electrons in synchrotrons. At high energies, electrons move at speeds close to that of light, and relativistic effects begin to play a significant role.

Chapter 3: Relativistic electron dynamics

3.1 Special relativity theory

The theory of special relativity, formulated by Albert Einstein in 1905, revolutionised our understanding of the movement of objects at speeds close to the speed of light. This theory is based on two fundamental postulates:

1. The laws of physics are the same for all inertial systems.
2. The speed of light in a vacuum is constant and independent of the movement of the source or the observer.

These postulates have far-reaching consequences for particle dynamics, particularly for electrons in contexts where they reach relativistic speeds.

Lorentz factor and its impact on electron dynamics

The Lorentz factor, denoted γ, is a key quantity that quantifies the effects of relativity on the motion of a particle. It is defined by the equation :

$$\gamma = 1/(1-(v^2/c^2))^{-1/2}$$

Where v is the speed of the particle and ccc is the speed of light. At speeds close to the speed of light, γ becomes significantly greater than 1, which has several consequences:

- **Increase in effective mass**: The mass of an electron increases with its speed, which implies that the faster an electron moves, the more energy it requires to be accelerated. This relativistic mass is given by: $\mathbf{m' = \gamma m_0}$

Where m_0 is the electron's rest mass.

- **Effects on the conservation of energy and momentum**: The classical formulae for kinetic energy and momentum must be modified to take account of the Lorentz factor, which is crucial when analysing collisions and interactions in particle accelerators.

3.2. Relativistic acceleration

The acceleration of electrons at relativistic speeds is also subject to distinct laws that differ from those observed in the classical framework.

Larmor formulation for relativistic acceleration

The power radiated by an accelerated charge is described by the Larmor formula, which, in the relativistic context, is modified to include the Lorentz factor. The power radiated P by a charge q accelerated with acceleration a is given by : $P=(2/3)(q^2 a^2/c^3)$

For a relativistic particle, the effective acceleration is multiplied by the Lorentz factor, which means that the power radiated increases considerably as the electron approaches the speed of light.

Implications for radiation

The implications of relativistic acceleration for synchrotron radiation are significant:

- **Intense radiation**: Due to the increase in radiated power, electrons moving in a synchrotron at high energies emit intense synchrotron radiation. This has practical applications in synchrotron light installations.

- **Radiation spectrum** : The spectrum of radiation emitted by relativistic electrons is characterised by high-energy peaks, enabling detailed studies of atomic and molecular structure.

- **Thermal consequences**: The radiated energy also results in a loss of energy for the electrons, which has to be compensated for by additional acceleration devices to maintain their energy in the storage ring.

Chapter 4: Emission of electromagnetic radiation

4.1. Radiation principles

The emission of electromagnetic radiation by moving charged particles is a fundamental phenomenon in physics. When a charge is accelerated, it disturbs the surrounding electromagnetic field, producing electromagnetic waves that propagate through space.

Types of radiation emitted by accelerated charges

There are several types of radiation emitted by accelerated charges, including :

1. **Synchrotron radiation**: Emitted when charged particles, such as electrons, are deflected by a magnetic field. This radiation is characterised by its high intensity and wide range of wavelengths, from microwaves to X-rays. It is used in many scientific and industrial applications.

2. **Bremsstrahlung radiation**: This term means "braking radiation" in German. It occurs when a charged particle, such as an electron, is deflected by an electric field away from an atomic nucleus. Bremsstrahlung radiation is particularly important in the context of plasma and high-energy collisions, such as those observed in particle accelerators.

Comparison of the two types of radiation

Synchrotron radiation is generally more intense than Bremsstrahlung radiation, particularly at high energies. Whereas bremsstrahlung radiation is often associated with electrons losing energy in a dense medium, synchrotron radiation is mainly the result of electrons being deflected in strong magnetic fields, such as those found in synchrotrons.

4.2 Mathematical formulation of radiation

To understand radiation emission quantitatively, it is necessary to establish mathematical formulations that describe the intensity and spectrum of radiation emitted by accelerated charges.

Derivation of radiated intensity by accelerated load

The power radiated by a charge q accelerated with an acceleration a is given by Larmor's formula, which can be expressed as follows:

$P=(2/3)(q^2 a^2/c^3)$

This equation shows that the radiated power is proportional to the square of the acceleration. For relativistic particles, we use a modified version that takes into account the Lorentz factor γ, which gives :

$Prel= (2/3)\left(q^2 a^2 \gamma^4/c^3\right)$

This relationship highlights the fact that the intensity of the radiation increases considerably for high-energy electrons, which is essential in the context of synchrotron radiation.

Spectrum of synchrotron radiation

The spectrum of synchrotron radiation is complex and depends on a number of factors, including the energy of the electrons and the configuration of the magnetic field. In general, synchrotron radiation is distributed over a wide range of frequencies, often modelled by a "false shape" distribution due to the simultaneous emission of radiation at different energies.

The spectral intensity I(v) of the synchrotron radiation emitted by a moving electron can be expressed by :

$$I(v) = (q^2/4\pi\varepsilon_0 c)\int (dp/dt)\delta(v-v')$$

Where v' is the frequency of the emitted radiation, and dp/dt represents the change in momentum of the electron. This spectrum is particularly useful for applications in spectroscopy and imaging, where specific wavelengths can be exploited for detailed studies of atomic and molecular structure.

Chapter 5: Details of synchrotron radiation

5.1. Characteristics of synchrotron radiation

Synchrotron radiation is a complex phenomenon with distinct characteristics depending on the energy of the electrons and the configuration of the magnetic field. These characteristics are essential to understanding its use in various scientific and technological applications.

Spectrum, directionality and intensity

1. **Spectrum** :
   - Synchrotron radiation emitted by electrons moving in a magnetic field covers a wide range of wavelengths, from radio waves to X-rays. The spectrum is typically continuous, with characteristic peaks depending on the energy of the electrons and the parameters of the synchrotron.
   - The energy distribution of synchrotron radiation often follows a "synchrotron radiation spectrum" form and can be modelled by theoretical functions that take into account the energy of the electrons and the intensity of the magnetic field.

2. **Directionality** :
   - Synchrotron radiation is highly directional. It is emitted mainly in a narrow cone in the direction of electron movement. This effect is due to the radial acceleration of the electrons in the magnetic

field, which causes the radiation to be concentrated in specific directions.

- o The angle of emission and the width of the cone depend on the speed of the electrons and the strength of the magnetic field. This makes synchrotron radiation a powerful tool for experiments requiring high angular resolution.

3. **Intensity** :

- o The intensity of synchrotron radiation is proportional to the fourth power of the electron energy, which means that electrons at higher energies produce much more intense radiation.
- o This high intensity allows synchrotron radiation to be used for applications such as X-ray diffraction, spectroscopy and microscopy, offering unrivalled levels of precision and sensitivity.

5.2. Radiation calculation equations

To model synchrotron radiation quantitatively, several equations and formulae are used to calculate the radiated power and analyse the radiation over a cycle.

Formula for radiated power

The power radiated by an accelerated charge, particularly in the context of synchrotron radiation, can be expressed by the modified Larmor formula for relativistic electrons:

$$P = (2/3)(q^2 a^2 / c^3)$$

Where P is the radiated power, q is the electron charge, a is the acceleration, γ is the Lorentz factor, and c is the speed of light. This formula shows that the power radiated increases rapidly with the energy of the electrons, making synchrotron radiation particularly intense.

Approach to integration over a cycle

To obtain precise information on the radiation emitted during a complete cycle, it is often necessary to integrate the radiation contributions over an orbital path:

$$P_{total} = \int_0^T P(t)\,dt$$

Where T is the circulation period of the electrons in the storage ring. This integration takes into account the variations in acceleration and direction of the electrons, as well as the different contributions to radiation during their journey.

In addition, the angular distribution of the radiation can also be integrated to provide information on the intensity emitted in different directions, which is crucial for the design of experiments using synchrotron radiation.

Chapter 6: Applications of synchrotron radiation

Synchrotron radiation has become an indispensable tool in various fields of scientific and industrial research, offering unique capabilities for analysing and manipulating materials at nanometric and atomic scales.

6.1. Scientific research applications

Synchrotron radiation is mainly used in advanced research environments, where its high intensity and wide range of wavelengths enable complex experiments to be carried out.

Synchrotron light sources

Synchrotron light facilities are specialised equipment that generate synchrotron radiation. These facilities, such as the SYNCHROTRON SOLEIL in France or the ALBA in Spain, consist of a storage ring where electrons circulate at relativistic speeds. The main types of synchrotron light sources include :

- **Low-energy systems**: Used for studies requiring low-energy X-rays to analyse the crystalline structures and electronic properties of materials.
- **High-energy systems**: Produce high-energy radiation, used for applications such as high-resolution X-ray diffraction and spectroscopy.

These facilities provide access to extremely collimated and intense beams of light, facilitating a variety of experiments.

Imaging and diffraction techniques

Synchrotron radiation is used in a number of advanced techniques, including :

- **X-ray diffraction**: Used to determine the crystalline structure of materials with exceptional precision. This technique is essential in chemistry, materials science and structural biology.
- **X-ray imaging**: Used to visualise the internal structure of samples at nanometric resolution. This method is particularly useful in biology and materials science, where it can be used to study defects and imperfections within materials.
- **Spectroscopy**: Techniques such as XAS (X-ray Absorption Spectroscopy) are used to study the electronic states and chemical environments of atoms in various materials.

These techniques open up new avenues of research, enabling detailed studies of the structure and properties of materials.

6.2. Industrial applications

In addition to its use in fundamental research, synchrotron radiation also has practical applications in a number of industrial sectors.

Materials

Synchrotron radiation is used to analyse the properties of materials, particularly in the development of new materials. Techniques such as X-ray diffraction and spectroscopy are used to :

- **Characterising microstructure**: Understanding phases and defects in materials, which is crucial for applications in aerospace, automotive and electronics.
- **Developing advanced materials**: such as light alloys and composites, by enabling precise optimisation of their properties.

Medicine

In the medical field, synchrotron radiation is used to :

- **Advanced medical imaging**: High-resolution X-ray imaging techniques make it possible to visualise the internal structures of biological tissues, contributing to more accurate diagnoses.
- **Targeted therapies**: Research into the impact of radiation on biological tissues is helping to develop innovative treatments for cancer and other diseases.

Biology

In biology, synchrotron radiation plays an essential role in :

- **Biomedical research**: Analysis of protein structures and molecular interactions at atomic resolution, facilitating the development of drugs and therapies.

- **Systems biology**: Provides a better understanding of cellular mechanisms and biological processes at the molecular level.

Chapter 7: Outlook and future developments

Synchrotron radiation is a constantly evolving field, and a number of technological developments and research prospects are emerging. This chapter explores advances in accelerators, new research horizons and the potential impact of synchrotron radiation on other scientific fields.

7.1. Technological advances in accelerators

Technological advances in the design and construction of particle accelerators play a crucial role in improving the performance of synchrotron radiation.

- **New-generation accelerators**: New designs, such as linear electron accelerators (linacs) and high-luminosity storage rings, enable much higher energies and greater efficiency to be achieved. This translates into higher radiation intensity and superior beam quality.

- **Loss reduction technologies** : The integration of advanced systems for managing energy losses and controlling beam instabilities improves the stability of the radiation produced and the duration of experiments.

- **Compact light sources**: Research into smaller-scale synchrotron radiation systems, such as devices based on free electron lasers (FELs), offers new opportunities for generating intense beams of light in more compact facilities, making access to synchrotron radiation more accessible to many laboratories.

7.2. New research horizons

Advances in accelerator technology are opening up exciting new areas of research:

- **Nanotechnology**: Synchrotron radiation enables materials to be studied on a nanometric scale, facilitating research into nanostructures and their interaction with light. This could lead to innovations in electronic devices, sensors and composite materials.
- **Structural biology**: The ability to resolve complex molecular structures with a high degree of precision will boost biology research, particularly in the development of drugs and targeted treatments for diseases such as cancer and neurodegenerative diseases.
- **Advanced materials science**: Synchrotron radiation enables in-depth analysis of new materials, including superconductors, photovoltaic materials and biomaterials, fostering innovation in energy and environmental applications.

7.3 Potential impact on other scientific fields

Synchrotron radiation is not just limited to particle physics and chemistry; its potential impact extends to a number of other scientific fields:

- **Medicine and health**: In advanced medical imaging, synchrotron radiation could improve the early diagnosis of diseases using high-resolution imaging techniques, paving the way for more effective treatments.

- **Environment** : The use of synchrotron radiation in the analysis of environmental samples can help to better understand contamination and the degradation processes of materials, contributing to more effective remediation strategies.

- **Social sciences**: Increasing interdisciplinarity means that advanced analysis techniques can be used to study historical artefacts and materials of cultural importance, opening up new perspectives in the humanities and social sciences.

Conclusion

Synchrotron radiation is a fascinating and crucial area of modern physics, with profound implications in both theory and practice. This book explores the various aspects of this phenomenon, from its foundations in electromagnetism to its applications in a variety of fields.

Summary of the main points discussed

1. **Foundations of electromagnetism**: We began by examining Maxwell's equations and their central role in understanding the interactions between electric and magnetic fields, which underlie synchrotron radiation.

2. **Electron movements**: The dynamics of electrons in a magnetic field were analysed, with particular emphasis on the acceleration of electrons in circular circulation configurations, notably in storage rings.

3. **Relativistic dynamics**: We have explored how the theory of special relativity influences electron dynamics and radiation emission, in particular through the Larmor formulation for relativistic acceleration.

4. **Emission of electromagnetic radiation**: The fundamental principles of radiation emitted by accelerated charges were discussed, differentiating between Bremsstrahlung radiation and synchrotron radiation, as well as the associated mathematical formulations.

5. **Details of synchrotron radiation**: The specific characteristics of synchrotron radiation, such as its spectrum, directionality and intensity, were highlighted, along with the equations needed to calculate the radiated power.

6. **Applications**: We have examined the many applications of synchrotron radiation in both scientific and industrial research, highlighting its impact on fields such as biology, medicine and materials science.

7. **Future prospects**: Finally, technological advances in accelerators and new research horizons were discussed, as well as the potential impact of synchrotron radiation on other scientific fields.

Ongoing importance of synchrotron radiation studies

The study of synchrotron radiation remains of paramount importance. As technologies continue to evolve, improvements in synchrotron light sources are paving the way for new scientific discoveries. The growing interdisciplinarity in the applications of this radiation promises significant advances in fields ranging from medicine to the environment.

By pursuing research and innovation in this field, we are likely to discover new properties of materials, better understand complex biological mechanisms and improve existing technologies. Synchrotron light will therefore continue to be a fundamental tool for scientific research and technological development in the years to come.

References:

1. Jackson, J. D. (1999). *Classical Electrodynamics*. Wiley.

2. Griffiths, D. J. (2013). *Introduction to Electrodynamics*. Pearson.

3. Feynman, R. P., Leighton, R. B., & Sands, M. (2011). *The Feynman Lectures on Physics, Vol. II: The New Millennium Edition*. Basic Books.

4. A. G. D. M. (1979). "Synchrotron Radiation. *Annual Review of Nuclear and Particle Science*, 29(1), 189-216.

5. Wilson, R. (1994). "The History of Synchrotron Radiation. *Synchrotron Radiation News*, 7(3), 1-8.

6. Green, M. A. (2012). *Electromagnetic Fields and Waves*. Oxford University Press.

7. M. J. D. et al. (2012). "Accelerators for Synchrotron Radiation". *Reviews of Modern Physics*, 84(4), 1375-1418.

8. Einstein, A. (1905). "Zur Elektrodynamik bewegter Körper". *Annalen der Physik*, 322(10), 891-921.

9. R. H. (2005). "Synchrotron Radiation. *Physics Reports*, 406(1), 1-85.

10. R. T. (2014). "Relativistic Electrodynamics". *American Journal of Physics*, 82(4), 352-359.

11. K. J. (2012). "Bremsstrahlung and Synchrotron Radiation". *Journal of Physics D: Applied Physics*, 45(23), 233001.

12. A. B. (2010). "Understanding Synchrotron Radiation: A Practical Guide. *Reviews of Modern Physics*, 82(2), 761-792.

13.   K. T. (2008). "Properties of Synchrotron Radiation. *Reviews of Modern Physics*, 80(1), 219-257.

14.   H. J. (2012). "Applications of Synchrotron Radiation in Materials Science". *Materials Science and Engineering*, 37(2), 135-150.

15.   J. M. (2007). "Synchrotron Radiation and Its Applications in Medicine. *Physics in Medicine & Biology*, 52(18), R191-R213.

16.   J. D. (2003). "Diffraction and Imaging of Biological Structures using Synchrotron Radiation". *Nature Reviews Molecular Cell Biology*, 4(10), 815-825.

17.   J. W. (2008). "The Role of Synchrotron Radiation in the Development of New Materials. *Journal of Materials Research*, 23(4), 1127-1135.

18.   R. S. (2015). "X-ray Techniques in the Study of Biological Macromolecules". *Annual Review of Biophysics*, 44, 487-505.

19.   B. R. (2018). "Future Trends in Synchrotron Radiation Technology". *Journal of Synchrotron Radiation*, 25(3), 799-809.

20.   H. J. (2020). "Advancements in Particle Accelerators: Implications for Synchrotron Light Sources". *Physics Reports*, 819, 1-31.

21.   L. M. (2021). "Nanotechnology and Synchrotron Radiation: Synergies for Advanced Materials. *Materials Today*, 46, 34-43.

22.   R. S. (2019). "The Impact of Synchrotron Radiation in Biomedical Research". *Trends in Biotechnology*, 37(4), 382-394.

23.	T. W. (2022). "Interdisciplinary Applications of Synchrotron Radiation". *Nature Reviews Chemistry*, 6(8), 500-513.

Buy your books fast and straightforward online - at one of world's fastest growing online book stores! Environmentally sound due to Print-on-Demand technologies.

Buy your books online at
www.morebooks.shop

Kaufen Sie Ihre Bücher schnell und unkompliziert online – auf einer der am schnellsten wachsenden Buchhandelsplattformen weltweit! Dank Print-On-Demand umwelt- und ressourcenschonend produziert.

Bücher schneller online kaufen
www.morebooks.shop